Project AIR FORCE

NATO'S FUTURE

IMPLICATIONS FOR U.S. MILITARY CAPABILITIES AND POSTURE

DAVID A. OCHMANEK

Prepared for the
UNITED STATES AIR FORCE

RAND

The research reported here was sponsored by the United States Air Force under Contract F49642-96-C-0001. Further information may be obtained from the Strategic Planning Division, Directorate of Plans, Hq USAF.

Library of Congress Cataloging-in-Publication Data

Ochmanek, David A.
    NATO's future : implications for U.S. military capabilities and
posture / David A. Ochmanek.
       p. cm.
    " MR-1162-AF."
    Includes bibliographical references.
    ISBN 0-8330-2809-X
    1. North Atlantic Treaty Organization. 2. United
States—Military policy. 3. Europe—Defenses. I. Title.

    UA646.3 .03 2000
    355' .031091821—dc21   99-086504

RAND is a nonprofit institution that helps improve policy and decisionmaking through research and analysis. RAND® is a registered trademark. RAND's publications do not necessarily reflect the opinions or policies of its research sponsors.

Published 2000 by RAND
1700 Main Street, P.O. Box 2138, Santa Monica, CA 90407-2138
1333 H St., N.W., Washington, D.C. 20005-4707
RAND URL: http://www.rand.org/
To order RAND documents or to obtain additional information,
contact Distribution Services: Telephone: (310) 451-7002;
Fax: (310) 451-6915; Internet: order@rand.org

Preface

The 1990s have been a decade of rapid change and adaptation for the North Atlantic Treaty Organization (NATO). As the Cold War came to a close and the Soviet Union collapsed, some called for an end to the alliance that had played a central role in countering Soviet military power for the previous 40 years. Others, impressed with the continuing need for transatlantic security cooperation, called upon NATO to adopt new missions, beyond its traditional role of territorial defense, aimed at meeting fresh security challenges.

Events have shown the need for a transformed NATO capable of undertaking a wide range of missions, including:

- Projecting stability into areas around the periphery of the NATO treaty area
- Intervening effectively in civil conflicts, such as those that have arisen in the former Republic of Yugoslavia
- Coordinating power-projection operations into areas such as the Persian Gulf region
- Countering weapons of mass destruction, both by impeding their proliferation and by preventing the use of such weapons.

An evolving consensus among the allies has led to a significant, if fitful, extension of NATO's mandate to encompass at least the first two of these missions. Over time, this expansion of NATO's area of regard and the scope of its missions is likely to continue, provided the Alliance's leading members have the political will to act when called upon to extend their influence outward. The questions that remain relate to the types of military capabilities most needed to make these missions a reality, and the rate and extent to which the United States and its allies are likely to field such capabilities. This report addresses these questions in the context of the Alliance's emerging strategy for advancing the interests of its members in peacetime, crisis, and war. It points to the need for a heavy emphasis on enhancing the deployability of NATO's military forces, and it forecasts a growing need for effective defenses against weapons of mass destruction and their means of delivery.

The research summarized here is part of a larger project on the implications of the changing strategic environment in and around Europe and its implications for the United States and NATO. The project, sponsored by the Commander-in-Chief, U.S. Air Forces in Europe, and by the Deputy Chief of Staff for Operations, Headquarters, United States Air Force, was conducted in the Strategy and Doctrine Program of RAND's Project AIR FORCE. This report should be of interest to those engaged in policy toward, or the study of, European security in the post–Cold War setting. Although its focus is on air forces and military units that can support air operations, its broad conclusions are relevant to all future U.S. and European forces.

PROJECT AIR FORCE

Project AIR FORCE, a division of RAND, is the Air Force federally funded research and development center (FFRDC) for studies and analyses. It provides the Air Force with independent analyses of policy alternatives affecting the development, employment, combat readiness, and support of current and future air and space forces. Research is carried out in four programs: Aerospace Force Development; Manpower, Personnel, and Training; Resource Management; and Strategy and Doctrine.

Contents

Summary

The United States and its allies face a host of challenges in the international sphere, including the looming threats of proliferation, regional conflict, terrorism, and the manifold problems associated with the failure of regimes to meet the needs of their populations. As daunting as these challenges are, veterans of the Cold War will be quick to point out that things could be worse. After all, NATO's member nations no longer face a plausible threat to their survival, as they did throughout the Cold War when Soviet military forces were deployed in the heart of Europe. Although some observers in recent years have bemoaned the loss of "certainty" and "predictability" that they (erroneously) think characterized the Cold War geopolitical situation, there should be no nostalgia for a period in which two nations with the power to destroy much of civilization pursued antagonistic security agendas.

The chief goal of the United States and its allies today, therefore, should be to preserve and consolidate an international situation in which no major power opposes or threatens their most important interests. They should pursue a security strategy whose core elements are intended to shape the behavior and expectations of key actors in ways favorable to the West's long-term interests. Of course, from time to time, NATO nations will have to employ their military and other assets to cope with challenges to their interests, and they will want to take steps to hedge against the possible emergence of serious new threats over the longer term. Across all three of these dimensions—shaping, coping, and hedging—the vast majority of challenges to the interests of NATO member states will arise from the periphery of the Alliance's enlarged treaty area and beyond.

There is no longer much debate about whether NATO should embrace new missions in addition to its traditional focus on territorial defense. Chief among the new missions are power projection (at least to the periphery of the treaty area, if not beyond), crisis management (or the ability to intervene effectively in smaller-scale conflicts), and countering weapons of mass destruction. For military planners, the key questions are: To what degree should their efforts focus on these new missions? What sorts of military capabilities are required to accomplish such missions? And what are the implications for U.S. and allied force planners? The following general findings emerge from our analysis:

- **The military forces of NATO's member states should be structured and postured for expeditionary operations.** Achieving a more expeditionary posture entails expanding and modernizing transportation fleets (principally military airlift, but also sealift), acquiring more mobile logistics assets, upgrading infrastructure in selected countries, and modernizing the forces themselves so that lighter, more mobile units can be more effective in a wide range of missions. This will entail, among other things, exploiting recent advances in surveillance, information processing, communications, and precision weapons so that the military assets of adversaries can be rapidly located, identified, and destroyed with minimal collateral damage.
- **The ability to deter and defeat chemical, biological, and nuclear weapons will become a growing preoccupation of the Alliance.** In addition to improving defenses against ballistic and cruise missiles, this implies that NATO should preserve the basic elements of the U.S. nuclear posture in Europe.
- **U.S. forces stationed in Europe are invaluable assets for shaping behaviors and expectations in the region and for responding to challenges in and around Europe.** The nature and locus of likely future challenges suggest that air bases in Italy and Turkey are particularly important strategic assets. Given the mix of peacetime and crisis response missions we foresee, the Department of Defense (DoD) should explore the idea of replacing at least one of the four U.S. heavy Army brigades in central Europe with a light infantry brigade, an air assault brigade, or some hybrid formation with greater strategic mobility and tactical flexibility.

The challenges of the future demand effective and coordinated action by nations with common interests. For the United States and its allies, NATO is by far the best vehicle available for organizing such action. NATO has made impressive strides toward adapting itself to the demands of a changed and changing world, but its members should recognize that neither the Alliance nor their own military establishments are yet perfectly suited to these demands. Fielding the requisite military capabilities will not come cheaply, nor will further institutional changes within NATO be cost-free. But it is well worth the effort.

Acknowledgments

This report represents, in part, a synthesis of research and analysis conducted by a team of RAND staff members who examined a wide range of challenges likely to face NATO members over the coming decade and beyond. As such, the author is indebted to his fellow team members, including Ian Lesser, who co-led the project, Tanya Charlick-Paley, F. Stephen Larrabee, Peter Ryan, Richard Sokolsky, Thomas Szayna, and Michele Zanini. The work was sponsored by General John Jumper, Commander of United States Air Forces in Europe (USAFE). His expansive and ambitious vision of NATO's future, and the American role in it, went far to animate and encourage all members of the research staff who worked on the project.

An early draft was reviewed by Ambassador Robert Hunter and by Marten van Heuven of RAND—both veteran diplomats with vast experience in U.S.–European security policy. Their ministrations improved the report greatly. RAND colleagues Zalmay Khalilzad and Robert Levine were also generous with their time and expertise, offering constructive suggestions at several points in the development of the report. The author also wishes to thank Major General Roger Brady, Major Kathleen Echiverri, and Major Jerry Gandy of USAFE Headquarters for their reviews of the report; and Colonel Joseph Wood, Commander of the 23rd Fighter Group, Colonel John Plant of Headquarters, United States European Command, and Major Phil Smith of Headquarters, United States Air Force, for their comments and suggestions. Jeanne Heller did her usual fine job of editing the text.

Chapter 1
The Emerging Security Environment

What is NATO? One veteran American diplomat, Marten van Heuven, has offered as good a definition as any. "NATO," he writes, "is a bundle of commitments, efforts, and procedures agreed to by a growing number of countries over the past half century to safeguard their vital interests."[1] This description of NATO emphasizes both its instrumental character and its institutional nature. As an instrument of security in a changed and changing world, NATO is what its members, by consensus, need it and wish it to be. Yet as an institution NATO is also rooted in past experience. As one considers the roles and functions that NATO could and should play in advancing the security of its members, it is important to keep in mind as well that human institutions can adapt only as quickly as the thinking and habits of their members change.

This report seeks to aid the process of adapting the Alliance to the demands and conditions of the future security environment, as best we can divine it. Accordingly, the report begins with a brief review of the main features of the international security environment as it bears on the interests of NATO's member states. It then examines the allies' (partly implicit) strategy for advancing their common interests in the face of new challenges and opportunities. It next offers implications for force planners in NATO and, in somewhat greater detail, the United States. It concludes with a review of the U.S. force posture in Europe and some remarks on the way ahead.

Baseball player and philosopher Yogi Berra is reputed to have observed, "Predictions are always dangerous; especially about the future." Those engaged in international security received a rude lesson in the risky art of futurology around a decade ago when, contrary to what most of us expected, the Soviet empire and then the Soviet Union itself

---

[1] Marten van Heuven, "NATO in 2010," unpublished paper presented to a conference sponsored by the Deutsche Gesellschaft Für Auswärtige Politik, Bonn, Germany, April 27, 1999, p. 1.

disintegrated—developments whose repercussions continue to alter the strategic landscape. The dangers of prediction notwithstanding, several fairly clear trends discernable now can help guide strategy and planning over the coming decade.

First, we can all but rule out the possibility that Russia will pose a significant, direct military threat to NATO over this period. This finding stands almost regardless of the eventual outcome of the social and political turmoil that the country is now experiencing. Modern military forces are expensive. They are also institutions that grow out of the societies whose interests they are meant to protect. Thus, a return to economic growth and a modicum of societal stability are necessary precursors to resurrected Russian military power. For this power to constitute a threat to the West, it would have to be wielded by a regime that viewed its interests as antithetical to our own, at least in some important respects. It is an open question whether these three requisites—economic success, social stability or mobilization, and anti-Western orientation—can once again coexist in post–Soviet Russia. However, it seems clear that they will not emerge there together and on a sustained basis any time soon.

Of course, Russia is the custodian of the world's second most powerful arsenal of nuclear weapons. For Russia, however, it seems that these weapons will be useful primarily as a guarantee against aggression rather than as an instrument of coercion. And while the United States and its allies will be compelled to devote substantially greater resources than now to capabilities for defeating weapons of mass destruction and their delivery vehicles, this imperative will be driven less by Russian nuclear capabilities and more by concerns over regional actors with expansionist or messianic objectives and anti-Western ideologies.

That said, a Russia ruled by elements that perceive reform and Westernization as somehow antithetical to their interests could cause significant problems for the United States and its NATO allies. Moscow could, for example, accelerate the spread of weapons of mass destruction and other sensitive military technologies through irresponsible export policies (or ineffectual export controls). Russia could also intervene in the politics of neighboring states in the Caucasus, Caspian, and

Central Asian regions in attempts to limit Western influence and undermine stability there. In addition, by wielding its veto in the United Nations Security Council or in other ways, Russia could play a spoiler's role, complicating or preventing efforts by the international community to promote common objectives through cooperative action.

The challenge for the West, then, is to try to keep Russia on a generally reformist path while continuing to support the independence and stability of Russia's neighboring states. Pending the emergence of a truly reformist, stable, and Western-oriented government in Moscow, this will require a careful balancing act, because some elements in Russia will continue to take a dim view of Western support to newly independent states on the Russian periphery.

Central Europe, by contrast, is becoming more stable. NATO's embrace of Poland, Hungary, and the Czech Republic is only one factor in this welcome development. Like several other states in the region (most notably, the Baltics), NATO's new members have energetically pursued reformist economic agendas within the framework of democratic political systems. The transition to open markets and pluralist politics has not been painless or risk-free for these countries, but the benefits are beginning to become manifest. The West's objectives here are to help sustain these trends by broadening and deepening the links (both informal and institutional) between these post-Communist success stories and the West.

Assuming a continuation of favorable trends in central Europe, the most pressing security challenges facing NATO will come from the Balkans and beyond Europe. The conflict over the status of Kosovo is only the most recent example of the ways in which the explosive cocktail of unreconciled ethnic or religious enmities, intermingled populations, and anti-democratic demagoguery can flare up into disputes that offend Western values and threaten Western interests. Put simply, large-scale, sustained, and Western-led intervention may still be required to help the peoples of the Balkans resolve the multiple conflicts dividing them. Montenegro, Macedonia, and Albania loom as potential flash points beside the tragedies that have been played out in Kosovo and that remain just beneath the surface in Bosnia. The major uncertainties relating to the Balkans are, first, the costs that Western governments will be willing to bear to try to stabilize the situation, and, second, the longevity of the Milosevic regime and the nature of the government that follows it.

The Balkans, in short, provide a test of the extent to which NATO's members are prepared to use their military forces (as well as their limited economic and diplomatic resources) in operations that aim to redress situations that threaten common yet rather amorphous interests. The Balkans are also a factor hastening the evolution of the Alliance toward a future in which out-of-area peace enforcement operations are seen—by its members and, increasingly, by the broader international community—as an accepted part of the Alliance's *raison d'etre*.[2]

Europe's southeastern periphery is another source of security concerns, especially because most of the military contingencies for which NATO actively plans are located in the southeast—especially around Turkey. The ongoing dispute between Greece and Turkey is just one of a number of possible conflicts that could involve Turkey, a country that shares borders with Syria, Iraq, and Iran, among others. Increasingly, the security of Europe (like that of the United States) is intertwined with situations in the Middle East and Eurasia. Iraq, Iran, Libya, and others have shown a determination to acquire weapons of mass destruction and the means to deliver them over increasing distances. At the same time, Europe and the West generally will remain heavily dependent on imported oil and gas, largely from the Persian Gulf. The growing ease and volume of international travel and trade are making it harder to protect against the spillover effects of violence in other nations or regions, whether these effects be "exported" deliberately by terrorists or inadvertently through large-scale migration by the victims of violence or state failure.

Further afield, the newly independent republics along Russia's southern border, as well as territories within Russia itself, are weak and unstable. These states face serious internal and intraregional threats to their stability and security, including regional, tribal, ethnic, and clan

---

[2] Not that many decisionmakers in NATO capitals are eager for another test similar to Operation Allied Force. Indeed, the searing experience of trying to coerce Milosevic to cease his depredations against the Kosovar Albanians will likely prompt great caution the next time such a challenge arises. Nevertheless, few question either that NATO was right to intervene when it did or that the allies must work to improve their capabilities for such military operations in the future.

disputes; economic hardships; immature political institutions, civil societies, and national identities; potential conflicts over land, water, energy, and mineral resources; and pervasive corruption, crime, and cronyism. In short, the Caucasian, Caspian, and Central Asian states generally suffer from the usual problems of weak states. Thus, the possibility cannot be ruled out that one or more of them could lose the capacity to maintain order and govern effectively. Although the West does not have vital interests in the region, it does have a stake in the success of reform there and in assured access to the region's resources. Hence, crises and instabilities there could engender pressure for intervention by NATO forces, either to monitor or to enforce a settlement of disputes.[3]

---

[3] For an analysis of NATO's interests in this region and challenges to them, see Richard Sokolsky and Tanya Charlick-Paley, *NATO and Caspian Security: A Mission Too Far?* RAND, MR-1074-AF, 1999.

Chapter 2
Changing Missions

Any list of security problems concerning the members of NATO will be dominated by challenges emanating from areas on the periphery of the Alliance's treaty area or, in many cases, well beyond it. There is little disagreement about this. What has yet to emerge is a clear picture of the roles NATO can and will play in dealing with this long list of challenges.

Not that the Alliance has stood still over the past decade. Two years after the fall of the Berlin Wall, NATO adopted for itself a new strategic concept to reflect these new realities. With this step, NATO, in the words of one commentator, "anointed itself as the guardian of European stability . . . reorienting its military forces away from border defense toward rapid reaction and power projection."[1] Arguably, the strategic concept adopted by NATO in 1991 should be regarded more as a reflection of the allies' aspirations than as a road map for force planning. Nevertheless, in the years since its adoption, NATO has taken steps to adjust its posture and activities toward the demands of new challenges—formally adopting peacekeeping as one of the missions of NATO's forces, for example, and opening a dialogue with selected countries bordering on the Mediterranean.

NATO's members, in short, are beginning to shift their conception of the Alliance away from one devoted primarily to the defense of territory to one increasingly focused on the defense of common interests. This shift in conception is occurring in fits and starts, and it is not shared evenly among the allies. But it seems an inescapable reality that the most serious threats facing NATO's members lie beyond the treaty area. Hence, NATO's "area of regard" is growing. The recent accession of three new members (and their almost immediate involvement in a

---

[1] Ronald D. Asmus, "Double Enlargement: Redefining the Atlantic Partnership After the Cold War," in David Gompert and Stephen Larrabee (eds.), *America and Europe: A Partnership for a New Era*, Cambridge University Press, New York, 1997, p. 37.

NATO military operation in the Balkans) is only one manifestation of this broader reality. Increasingly, the members of NATO will seek to influence events beyond the NATO treaty area. It is surely in the interests of the United States to encourage its allies to see NATO as a vehicle for extending this influence.[2]

What might be called this *"Drang nach Aussen"* underlies the new strategic concept that the Alliance formally adopted in 1999. That concept recognizes the need for the Alliance to take account of risks and threats beyond those related to the defense of territory. These beget demands for capabilities to carry out "new missions:"

- Power projection—the ability to deploy and employ military forces and assets rapidly, over long distances, and for sustained periods
- Crisis management—the ability to intervene diplomatically and militarily in disputes that could result in conflicts ranging from small scale to fairly large
- Countering weapons of mass destruction—reducing incentives by others to acquire or employ weapons of mass destruction and preparing to prevent the effective use of such weapons
- Countering terrorism.

These changes in thinking about the purposes of the Alliance and the primary focus of its efforts are welcome. However, as important as a general commitment to adopt new missions is, such commitments are, in the end, meaningful only if they lead to actions over a sustained period to implement them. Key questions remain:

- What concrete responsibilities and operations will NATO and its members undertake to make these general missions a reality? Where will they draw the line between matters that are the business of the Alliance and those that are to be left for unilateral action or responses by *ad hoc* coalitions of the willing?
- What sorts of military capabilities—hardware, trained people, operational concepts, and supporting assets—and activities are most appropriate for carrying out important new missions?

---

[2] For a range of views regarding NATO's future roles and of the alliance's place in U.S. global strategy, see Gompert and Larrabee, 1999.

- To what extent are the members of the Alliance fielding the military capabilities needed to undertake these missions? How interoperable are their forces?
- And, for the United States, how should its military forces in Europe be shaped in light of future challenges and strategies?

The remainder of this report addresses these questions, offering first an overall strategy that could govern NATO's activities in peacetime, crisis, and conflict. It then suggests some broad implications of this strategy for force planners in NATO member countries, focusing finally on some more-specific implications for U.S. planners.

Chapter 3
Objectives and Strategy

Our review of trends affecting Western interests portrays an international security environment that looks fairly benign over the near to mid term, at least in contrast to the situation that prevailed in Central Europe through the Cold War. The present is certainly not free from challenges, however, as the situations in the Balkans, the Middle East, the Gulf, Korea, Taiwan, and elsewhere highlight. In addition, more serious threats loom on the horizon in the forms of proliferation, regional conflict, terrorism, societal unrest, and the chaos that results from the failure of states to provide order within their borders and to meet the basic needs of their populations. It seems self-evident that these threats can affect the security and well-being of citizens in all of the allied countries, in some cases profoundly. Hence, we have a shared interest in preventing such threats to the degree we can.

All of this can be distilled into a single, overarching statement of objectives for the United States and its allies: *The chief security objective for the Western allies today is to shape behaviors and expectations so that serious new threats do not emerge. A secondary priority is to contain and defuse existing threats.*

These objectives suggest the need for a two-tiered strategy:

- The core strategy for the United States and its allies should be focused on shaping the expectations and behaviors of key actors in the international security environment so that key regions of the world evolve in ways favorable to the allies' long-term interests. This will entail, *inter alia*, acting when appropriate to broker or enforce resolutions to conflicts that affect U.S. and allied interests. A shorthand for this core strategy might be *shaping and coping*.

- This core strategy should be complemented by efforts to deal with the possibility that the United States and its allies will not be universally successful in preventing the emergence of serious new threats. Hence, a *hedging* strategy is needed to ensure that the requisite capabilities will be in place to meet more stressing threats that might arise.

What do these strategies mean in specific terms?

SHAPING

The United States and its allies bring to bear a host of instruments—many of them nonmilitary—in their efforts to shape the behavior and expectations of others. As noted previously, formally extending security guarantees to selected states is perhaps the most powerful tool available to policymakers interested in stabilizing regions where important interests are at stake. The extension of security guarantees reduces pressures on new members to seek costly and potentially destabilizing military capabilities that might be seen as needed should a newly independent state have to provide for its own defense. Prospective membership in NATO has also had notable side effects, including increased investor confidence and the accelerated economic growth that accompanies investment capital. And by outlining criteria desirable for new members, including the renunciation of force to settle disputes, adherence to democratic principles, and respect for minority and human rights, NATO has helped to promote adherence to responsible and stabilizing policies. Indeed, the possibility of accession into NATO has provided strong support to those who seek to promote reformist agendas in states formerly ruled by communist parties.

The military forces and staffs of NATO member states have promoted the spread of Western-oriented systems and policies by providing advisory assistance to their counterpart organizations in countries formerly dominated by Moscow. Since the early 1990s, U.S. and allied military establishments have been sending teams of specialists—commissioned and noncommissioned officers, as well as civilians—to foreign defense ministries and parliaments where they instruct their hosts on subjects ranging from NATO military doctrine and training to defense planning and programming. These advisory teams also provide examples of Western-style civil-military relations, emphasizing the importance of the rule of law and the roles of elected legislatures and executives in controlling military expenditures and activities. Participation by officers of other countries in NATO nations' professional military education courses reinforces these messages.

Multinational field training, exercises, and operations have always been important vehicles within NATO for improving military proficiency and increasing the ability of forces from different nations to operate jointly. Since the early 1990s, NATO has conducted an increasingly vigorous program of combined training exercises involving non-NATO nations, as part of the Partnership for Peace (PFP) program. These events reinforce messages given by Western military advisors. They also are invaluable for increasing mutual trust and confidence among people from nations that were for decades locked in an adversarial relationship. And when U.S. or allied forces deploy to another country to conduct training, this sends a message to friends and potential foes alike that Western forces have the capacity to contribute to that country's defense.

Shaping activities are also directed toward reducing the threat posed by weapons of greatest concern. In particular, containing the threat posed by longer-range ballistic and cruise missiles and by weapons of mass destruction (WMD)—nuclear, chemical, and biological weapons—will become a growing preoccupation of the United States and its allies in the years to come. In the hands of states bent on challenging Western interests, these weapons can pose serious challenges to forces operating within their range. In the hands of terrorists, they raise the prospect that small groups that are difficult to identify, deter, or strike can inflict horrific damage on defenseless populations.

NATO has reduced proliferation pressures in Europe in the first instance by helping to stabilize the region. During the Cold War, NATO's security umbrella (centered on American security guarantees and supported by nuclear and conventional arsenals) allowed most of the technologically advanced nations of Western Europe safely to forgo the nuclear option. The same is true for NATO's new members, some of whom might have felt compelled to move toward the development of independent deterrent forces in the absence of credible security guarantees. Similarly, NATO's less-formal assurances to Ukraine helped that nation to rid itself of nuclear weapons and delivery vehicles left there after the demise of the Soviet Union.

In spite of these successes, the prospect is for growing missile and WMD threats from states on Europe's periphery. NATO nations will require a multipronged approach to the problem. Specifically, the United States and its allies will seek to:

- Impede the flow of technologies related to the development of WMD, ballistic missiles, and cruise missiles to nations and groups espousing hostile security objectives
- Deter the use of WMD by retaining the capacity to retaliate with devastating effects
- Defeat attacks on Western and allied forces and territory through a combination of active defenses, passive protection, and capabilities to locate and destroy WMD before they are used
- Develop operational concepts for power projection that reduce the exposure of expeditionary forces to WMD attacks.

We return to the implications of the WMD problem later, but suffice it to say for now that neither NATO nor its members have yet come fully to grips with the problem in their defense planning and resource allocation priorities.

COPING

The coping dimension of the core strategy involves efforts to resolve or neutralize existing and emerging threats to Western interests. In the post–Cold War period, NATO members have faced immediate challenges from overt aggression by Iraq in 1990, as well as from terrorist attacks, simmering disputes between Greece and Turkey and within Turkey itself, and, most wrenchingly perhaps, persistent and widespread violence in the former Yugoslavia. Conflicts further afield, including the genocide in Rwanda and the rampant violence in Algeria, have also raised concerns among populations and governments in NATO member countries.

Often military power must be brought to bear, either implicitly or explicitly, to compel a satisfactory resolution to such challenges. Depending on the circumstances, virtually the full range of military capabilities may called upon. Capabilities often in demand include:

- Multinational planning, coordination, and control of military operations
- Providing humanitarian relief to victims of violence or natural disasters
- Monitoring military activities in regions where conflict might occur or is occurring
- Imposing and monitoring embargoes on the shipment of unauthorized goods by land, sea, or air
- Training and equipping local forces for their own defense
- Rapidly deploying air, naval, land, and amphibious forces to regions in conflict, and sustaining these forces once deployed
- Coercing or punishing enemy leaders through precise attacks on military, political, and economic assets
- Preventing or defeating military actions by one or both parties to a dispute.

The latter may call for capabilities to accomplish a wide range of operational objectives, including protecting civilian populations from attacks by armed factions; gaining air and naval superiority; delaying, damaging, and destroying light and mechanized ground forces, supply columns, and their support infrastructure; providing support to friendly ground forces in the form of supplies, transportation, information, or fires; and defeating enemy ground forces in battle.

This list of military capabilities relevant to NATO's coping tasks will be familiar to students of modern military operations of almost any time and place. It bears noting, however, that the context and conditions under which NATO's forces may be operating these days are likely to be quite different from the conditions for which those forces prepared a decade ago. These changes are summarized in the figure on the next page. During the Cold War, NATO's planning and training were dominated by concerns about deterring and defeating large-scale aggression. In the event a major war in Europe occurred, it was as-

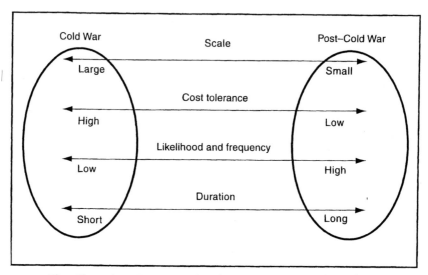

The Changing Context of Military Operations by NATO

sumed that millions of combatants would be engaged and that truly vital national interests would be at stake, including the very survival of NATO's member nations and their populations. Under these circumstances, the United States and its allies were willing to risk high costs in the event of war, including the loss of many thousands of combatants and the incalculable losses that would result from a large-scale nuclear exchange. The only solace one could take from this situation was that the potential costs and risks for both sides were so astronomical that the likelihood of war in Europe was deemed to be quite low.

Today, the interests that NATO's members have at stake in most prospective conflicts—such as violence in the Balkans—are, relative to "the big one," modest and ambiguous. The territory, lives, and prosperity of most NATO nations are not today directly at risk from any plausible conflict (Turkey and perhaps Greece being the sole exceptions). Yet the interests involved in areas around NATO's periphery are not negligible: Western nations have important economic interests in the Persian Gulf and Mediterranean regions. Even where no readily definable major national interests exist, many people in Europe and America feel some responsibility for the well-being of others, at least to the level of enforcing basic norms prohibiting officially sponsored violence against innocent civilians. Such concerns as these are prompting the leading nations of the international community to accept, however haltingly, the proposition that, in order to preserve international stability and the well-being of all, they must do what they can to prevent the most egregious violations of human rights in "local" or internal conflicts.

By and large, citizens of NATO's member states feel these responsibilities most keenly with regard to societies within Europe itself. This is understandable, given that violence in neighboring states can readily have spillover effects at home and that feelings of kinship are naturally stronger among groups with a shared historical and cultural background.

The scale of conflicts for which we must prepare today ranges from quite small (a battalion or so, widely scattered, was deemed sufficient to deter deliberate aggression against Macedonia for several years prior to NATO's intervention in Kosovo) to moderate in the case of another war in the Gulf. Appropriately, given the level of interests at stake, tolerance for costs—most especially human costs, in terms of casualties—is low. At the same time, these conflicts arise with depressing frequency and their underlying causes may persist for years, demanding the commitment of outside forces for extended periods.

This very different context for post–Cold War contingencies has profound implications for NATO strategists and force planners. It means, among other things, that for military instruments to be relevant to the problems with which the Alliance is now coping, those instruments must be capable of being employed with high confidence that they can achieve their objectives without risking disproportionate loss of life. Not only will high casualties not be tolerated among friendly combatants, it has become increasingly clear that our forces must not, through their operations, make the situation appreciably worse for civilians in the region. That is, "collateral damage" must be kept to a minimum. "Technology," observes one seasoned European strategist, "is not a luxury . . . the greater [our] concern about casualties, the greater the reason to exploit our technological advantages."[1] This is a

---

[1] Air Vice Marshal Professor Tony Mason, "The Future of Air Power: Concepts of Operation," unpublished paper delivered November 28, 1997 in The Hague, Netherlands.

demanding set of criteria that cannot always be met, particularly when the forces we are trying to defeat adopt tactics that make it difficult to locate combatants and to separate them from the populace. But Western defense planners can anticipate that their nations' decisionmakers will evince an enduring interest in military capabilities that can be employed with minimal risk of losses and collateral damage.

HEDGING

Hedging involves taking prudent and affordable steps to prepare for unlikely but threatening future contingencies. If preparing to meet the threats of WMD and small- to moderate-scale military operations is an integral part of NATO's core strategy, the principal unexpected threat against which NATO must plan would seem to be the potential reemergence of a large-scale conventional and nuclear military threat on or near NATO's borders. As noted previously, the odds of such a threat emerging from Russia in the near term or even the mid term seem long indeed. This suggests that, whatever measures are deemed prudent for hedging against this threat, no urgent steps are called for. Nevertheless, fielding the capabilities needed to cope adequately with an adversary that deployed even one-third of the forces that the Soviet Union could have brought to bear against NATO in the 1980s would be a costly and time-consuming task. For this reason, prudence demands that steady efforts be made over the coming years to lay the foundations for such capabilities.

Much of the necessary work will be undertaken as a matter of course: Plans are afoot to improve the readiness, training, equipment, and interoperability of NATO's forces so that they may be able to carry out the missions assigned to them under the core strategy. For the most part, appropriate additional preparations center on developing infrastructure suitable for supporting large-scale military operations, particularly to the east and south of NATO's enlarged treaty area. This infrastructure encompasses both the military and civil spheres. For example, military command and control facilities and communications centers in both old and new member states should be upgraded. Selected air bases in new member states also require upgrading. On the civil side, transportation networks (roads, bridges, rail lines, and air traffic control), petroleum distribution networks, communications, and power grids all are relevant to military operations and may be in need of modernization. Again, most of these enhancements will be undertaken in any case as part of broader efforts to modernize the economies of new member states and to provide for the effective control of national borders and airspace.

One special element of the hedging strategy is the infrastructure associated with U.S. nuclear weapons in Europe. All three dimensions of the strategy outlined here point to the value of retaining in Europe the capability to retaliate against those who might employ weapons of mass destruction against U.S. or allied assets in or around Europe. Such a capability is a useful hedge against the possible reemergence of a major military threat on NATO's borders. It also helps to shape the environment positively by reducing the incentives others might have to acquire WMD or to threaten to unleash them against Europe. And, should such weapons be used, having the means of retaliation deployed in Europe would help to ensure a consultative and decision-making process among the allies characterized by shared risks and burdens. For these reasons, the Alliance will want to preserve the basic elements of the currently deployed capability.

Chapter 4
Implications for NATO's Force Planning

This examination of the security environment, NATO nations' objectives, and a strategy for advancing those objectives suggests some clear implications for force planners. First, the overwhelming majority of NATO's military activities in the coming years will take place at a considerable remove from the home stations of most forces. The most likely NATO shaping and coping operations, such as conducting joint exercises with PfP states, monitoring compliance with international agreements, or helping to enforce peace in the Balkans, will take place beyond even the enlarged NATO treaty area. Even potential Article V operations—coming to the aid of a NATO member state, such as Turkey, for example, that might be the target of aggression—will be of an expeditionary nature for most of NATO's forces.

This assessment suggests strongly that the forces most useful for NATO will be those that are capable of rapid deployment and are postured for expeditionary operations. This means more than buying additional military transportation assets, although these are indispensable to force mobility. It means investing in logistics and support assets that are either forward based or can move as rapidly as the forces they support. Air bases in and around NATO's new member states should be upgraded as well. We have seen, for example, that air bases in Hungary can support fighter aircraft operations over the former Yugoslavia. Likewise, bases in eastern Romania or Bulgaria could support fighter operations over the Black Sea, Turkey, and Ukraine. The value of forward operating bases such as these is greatly magnified if preparations have been made in advance to support high-tempo operations. Depending on the base, such preparations might entail repairs to runways or facilities, enhancements to fuel storage and pumping capabilities, prepositioning of ground support equipment and munitions, and improvements to the physical security of the facility.

Ground forces, too, need to have access to a wide array of support assets to sustain operations away from home. They can be transported with the units or, conceivably, prepositioned in areas of likely deployment. Either way, the costs of posturing forces for true expeditionary operations can be significant, independent of the price of their transportation assets.

As planners in NATO countries consider ways to adapt their forces to these demands, they would do well not to consider existing operational concepts and force structures as given. As some elements of the coalition forces' campaign in the Gulf War showed and the air campaign over Kosovo confirmed, military operations can be transformed through a combination of new capabilities for surveying activities on the battlefield, dynamically controlling military operations, and engaging and attacking targets with guided weapons. In particular, it appears that lighter, more-mobile forces can, by exploiting advances in information and firepower, accomplish key operational tasks that were previously the primary domain of heavy, armored forces.

Aircraft and artillery, for example, have traditionally been regarded as useful in disrupting, delaying, or "softening up" enemy ground forces. As such, their role was to support friendly ground forces whose job it was to administer the *coup de grace* in a close battle. Today, modern reconnaissance, communication, and computing capabilities, coupled with precision munitions, are allowing modern militaries to engage and destroy mobile ground forces at unprecedented levels of effectiveness with airpower and longer-range fires. Over time, the adoption of novel operational concepts along these lines will shift the division of labor on many battlefields to forces that may be more easily adapted to the demands of expeditionary operations than a traditional mix of heavy, mechanized ground forces and supporting fires.[1] Hence, the modernization of NATO's military forces, which is needed to meet the demands of a new strategy and a changing threat environment, may have the added benefit of facilitating the transition of NATO's forces to a more expeditionary posture.

---

[1] For an analysis of the potential of modernized joint forces to defeat an armored invasion, see David Ochmanek et al., *To Find and Not to Yield: How Advances in Information and Firepower Can Transform Theater Warfare*, RAND, MR-958-AF, 1998.

NATO nations have adopted a broad-based Defense Capabilities Initiative (DCI), intended to accelerate the fielding of military capabilities best suited to the new environment and challenges NATO faces. The primary objectives of the DCI are to provide forces that are more capable of effective expeditionary operations outside of the treaty area and that are more interoperable with those of other NATO nations. Special emphasis is being placed on enhancements to capabilities for collecting, processing, and exploiting information; accelerated procurement of advanced munitions; and improved capabilities for operating in environments that may be contaminated by weapons of mass destruction. The DCI seeks to stimulate progress in five key areas: deployability, sustainability, effective engagement, survivability, and communications.

Neither extensive force modernization nor the purchase of new transportation and mobile logistics assets will come cheaply. NATO's European members are planning to make gradual improvements, but progress will be uneven and many nations will find themselves unable to modernize their forces rapidly or extensively. A gap exists, therefore, between the forward-looking rhetoric of NATO's new strategic concept and the capabilities of many of NATO's forces to support the missions inherent in that concept. This gap will remain for some time to come, even in a best-case scenario. Allies can contribute to combined operations, in other ways, however—for example, by taking steps to facilitate operations by deployed forces on or through their territories. Assured access to en route bases in central and southern Europe is crucial to U.S. airlift operations in the eastern Mediterranean and Persian Gulf regions. By upgrading facilities at selected airfields, such countries as Poland, Romania, and Turkey can help ensure prompt and effective operations by NATO forces during peacetime, crisis, or conflict.

Chapter 5
Implications for U.S. Force Planners

All three dimensions of the strategy for NATO outlined here have implications for U.S. force planners. In many ways, these implications suggest the need for change at the margins. This is, in part, an accident of history: The United States has long had the good fortune of being able to fight its wars far from home. Particularly since World War II, U.S. forces have been postured and equipped for operations in Europe and Asia, regions that remain the major foci of U.S. national interests and power projection. Thus, the military forces of the United States already possess large airlift and sealift fleets, forces configured for expeditionary operations, and a network of overseas alliances and bases to support operations abroad. Nevertheless, as potential adversaries acquire new and more threatening capabilities, more will need to be done to provide the basis for a prompt and effective riposte in the face of short-warning aggression.

SHAPING

The U.S. Navy and Marine Corps have long stated that providing an "overseas presence" is an important part of the missions assigned to their forces. In fact, forces from each of the U.S. military services that are stationed or deployed abroad provide "presence." The presence of U.S. forces (and their families) in Europe and other regions during the Cold War was vital to underscoring the credibility of U.S. security commitments. U.S. forces based abroad have also played indispensable roles by providing a basis for an immediate response to aggression and by participating in joint and combined training exercises with the forces of allied and friendly nations.

That said, the end of the Cold War has led to different and, in some ways, more-demanding requirements for these and related activities aimed at promoting stability and security ties with the West. Even as these demands have grown, the number of U.S. military forces and personnel in Europe has fallen. Because of this, the United States today is missing valuable opportunities to interact with the forces and planning staffs of military establishments formerly under the sway of the Soviet Union. Although the potential demand for such interactions is, of course, virtually limitless, in light of limits on resources and competing demands, we will have to content ourselves with conducting only the highest-priority "influence projection" missions.

To enhance the quality of these interactions with counterparts in allied and other militaries, the Secretary of Defense in 1997 directed each of the services to create programs similar to the U.S. Army's Foreign Area Officer (FAO) career specialty. Accordingly, the Air Force has begun to designate officers with fluency in a foreign language and special knowledge of designated countries as FAOs; and the service is providing incentives and opportunities for more officers to acquire such skills through formal education and experience in the field. All told, the Air Force anticipates building a corps of 500 to 700 FAOs. Plans also call for increasing the number of Air Force officers proficient in foreign languages to 6900 (approximately 10 percent of all active duty and reserve officers) by 2005. As these officers gain increased knowledge of other countries and develop personal contacts there, the value of ongoing "outreach" efforts, already high, will grow substantially.

Another low-cost, high-payoff means for projecting influence is the practice of inviting military officers, noncommissioned officers (NCOs), and government officials from foreign countries to attend professional military education courses in the United States and other allied countries. Courses at the National, Army, Naval, and Air War Colleges; the Army and Air Force Command and General Staff Colleges; the Industrial College of the Armed Forces, and others do more than teach Western military planning and doctrine. They also create unique opportunities to form personal bonds that often last through the careers of the graduates and greatly facilitate international cooperation and understanding. Planners in DoD should identify key countries where they would like to promote accelerated Westernization and reform and then provide the military services with the (modest) additional resources that would be needed to increase the number of students from these countries in their most prestigious schools.[2]

The quality of multinational training exercises also can be enhanced. Reports from U.S. Air Force units involved in recent NATO training exercises suggest that allied, partner, and U.S. forces alike would get more out of these efforts if clearer guidance were provided to all participants regarding the desired focus and objectives of the training. Likewise, more effort should be invested in distilling, assessing, and promulgating operational, tactical, and procedural lessons learned from each exercise. Finally, units could benefit from having more time to prepare for exercises prior to the actual deployment.[3]

The number of multinational exercises NATO can support in a year will always be limited: Few soldiers, sailors, or airmen would wish to be deployed away from their home station more than they have been in recent years. When thinking about how to get the most out of our forces for a given level of operations tempo, two points stand out: First, there is simply no substitute for forward basing of U.S. forces. The difficulties and costs associated with deploying personnel from a Europe-based unit to a NATO exercise, while not trivial, are far less than those associated with deploying the same assets all the way from the United States. It is all but inevitable that a reduction of U.S. forces from Europe would lead to a proportionate reduction in U.S. military outreach activities there—with negative consequences for our core strategy and U.S. influence.

Second, it may be possible to manage routine operations so as to achieve a modest increase in interactions with selected military forces from new member or partner nations. One approach would be to begin shifting some routine training by U.S. air and ground forces to training areas on or near the territory of new NATO members and PfP nations. By doing so, U.S. and other forces from Western countries could combine some of their essential routine training with military outreach, inviting host country forces to observe and accompany de-

---

[2] Such an initiative would require, for some countries, an accompanying increase in English language training resources as well.

[3] It is not unusual for U.S. Air Force squadrons in Europe to receive their "operations orders" to deploy to a NATO exercise only a day or two prior to execution.

ployed units on training missions. Because high-quality training frequently demands special support facilities (such as live-fire and instrumented ranges), creating a suitable training environment in new areas might involve significant costs. However, over time, it may make sense to move in this direction. A similar approach could be taken with respect to "real world" operations. For example, officers and NCOs from new member countries could spend time observing and, increasingly, participating in the planning and execution of such operations as the patrolling of the skies over Bosnia or northern and southern Iraq.

COPING

The implications of the coping dimension of the core strategy are essentially the same for U.S. forces as for their allied counterparts. First, U.S. forces, whether based abroad or at home, will need to continue to enhance their ability to deploy rapidly and to operate effectively from deployed locations. Continued modernization of the U.S. Air Force's fleet of military airlift aircraft is called for. Some units—notably, elements of the bomber force—also require additional support equipment so that they can quickly deploy and operate away from home station. The U.S. Army is also taking some steps to enable its forces to deploy more quickly. To date, these have primarily involved prepositioning heavy equipment and supplies in areas of potential conflict. Over the past several years, for example, the Army has substantially increased stocks of equipment in the Persian Gulf region.

Second, it will become increasingly important for U.S. forces to be able to operate effectively in the face of enemy capabilities to deliver weapons of mass destruction, and to defeat such weapons before they can strike their targets. The main implications of this have been touched on above. In addition to deploying more-effective defenses against ballistic and cruise missiles, U.S. forces will need to get better at finding and destroying small mobile targets (notably, mobile missile launchers and their associated vehicles) and deeply buried targets where WMD might be stored. Finally, it is worth repeating that the basic elements of the U.S. nuclear posture in Europe should be retained in anticipation of the day when all of Europe falls within range of missiles and WMD from any of a number of hostile or potentially hostile states. Deterrence through the threat of retaliation will be a necessary if not sufficient answer to these emerging threats, and it seems self-evident that NATO's threats to retaliate will be more credible if the means of retaliation are deployed in Europe and subject to NATO release authority.

Finally, U.S. and allied military forces should evolve toward greater effectiveness in the difficult environment of smaller-scale operations. The earlier discussion in Chapter Three highlighted the fact that the operations most frequently undertaken by NATO forces will be characterized by the need for highly asymmetric outcomes. That is, U.S. and allied forces will be expected to achieve their objectives fairly quickly while risking few losses and inflicting minimal unwanted casualties. Such objectives might range from monitoring military activities in a distant country, to coercing recalcitrant leaders, to destroying an enemy's fielded forces.

These are demanding criteria, and in some cases military planners will be unable to assure their civilian leaders of success. But to the extent that Western militaries can improve their capabilities for rapid mobility, all-weather/day-night reconnaissance, suppression of enemy air defenses, precision attack, and force protection, they will become increasingly relevant to the needs of policymakers struggling to protect Western interests in a turbulent world.

HEDGING

The main implications of NATO's hedging strategy for U.S. force planners are the same as those mentioned above. Principally, U.S. defense officials and military leaders should expect that they and their allied counterparts will be compelled increasingly to deal with the possibility that a hostile state or group could threaten allied territory, populations, and forces with weapons of mass destruction and delivery vehicles of growing range and sophistication. This calls for accelerating investments in active warning and defense systems. It will also mean retaining, on European soil, the means of retaliating against such attacks.

Measures that would be useful in deterring or defeating a future large-scale conventional military threat in Europe—the other concern animating the hedging dimension of the strategy—are, by and large, those that would also help ready U.S. and allied military forces for effective expeditionary operations. In addition to the modernization priorities outlined above, air bases should be upgraded along the periphery of the treaty area.

Chapter 6
Some Thoughts on U.S. Forces in Europe

This report has offered the view that the United States should continue to station and deploy substantial numbers of capable military forces in Europe. In a region where the shaping of behaviors and expectations lies at the heart of our security strategy, military forces must sustain a high level of engagement with allies and friends. In light of these considerations, there is no compelling reason to reduce overall U.S. force levels in Europe; indeed, the arguments against such reductions are numerous. This should not be taken to mean, however, that the currently deployed mix and basing of U.S. forces ought to be immutable: It may be that U.S. forces in Europe today are not optimally structured or postured for the tasks they must perform. A full assessment of this question lies beyond the scope of this study, but several guideposts for policy emerge:[1]

- Manpower at major headquarters should be maintained at levels sufficient to sustain at least the current pace of advisory assistance activities under way with new NATO members and PfP states. Efforts should be made to develop a growing cadre of officers and NCOs with language abilities and other skills called for in "military diplomacy" missions.
- Forces stationed in Europe should be capable of the types of operations most commonly practiced in multinational exercises and relevant to prospective combined operations in and around Europe. These include the deployment of combined joint task force (CJTP) headquarters and communications nets, all-weather/day-night surveillance, basic infantry maneuver tactics

---

[1] For a more thorough assessment of needs and opportunities relating to U.S. forces in Europe and elsewhere overseas, see Richard L. Kugler, *Changes Ahead: Future Directions for the U.S. Overseas Military Presence*, RAND, MR-956, 1998.

(up to battalion level), small- and large-scale gunnery, air defense (including dissimilar air combat tactics—DACT), suppression of enemy air defenses (SEAD), ground attack, and base security. NATO forces also might begin to place increasing emphasis on military operations in urban terrain (MOUT).

- U.S. forces in Europe should be capable of rapidly deploying to new operating locales and conducting high-tempo operations.

Taken together, these general findings suggest that the U.S. Air Force presence in Europe is well-balanced (albeit modestly sized) and suited to its missions in peacetime and conflict. The forces consist of one squadron of ground attack aircraft (A/OA-10s), two squadrons of air superiority aircraft (F-15Cs), two squadrons of aircraft specialized for surface-to-air missile (SAM) suppression (F-16CJs), and four squadrons of multirole F-15E and F-16 aircraft.[2] These combat fighters are supported by C-130 airlifters, KC-135 aerial refueling aircraft, and other support aircraft. Although these assets must generally be supplemented by reconnaissance and other specialized aircraft for the conduct of most operations, taken together they offer commanders in Europe a fairly complete and flexible package of air warfare capabilities relevant to a wide range of contingencies.

The fit between U.S. Army forces stationed in Europe and their missions is somewhat less clear-cut. The Army has four heavy armored and mechanized infantry brigades stationed in Europe as part of two understrength divisions. These forces constitute an important component of the U.S. military posture in Europe and provide essential capabilities for such missions as the Bosnian peace implementation force (IFOR) and the NATO-led peacekeeping force in Kosovo (KFOR). However, many of the armies in eastern and central Europe are structured around a mix of light and heavy forces. A similar mix of U.S. Army forces in Europe would seem best-suited to supporting combined training exercises. Considering as well the premium placed on rapid mobility and the growing demand for forces to do such "constabulary" operations as peacekeeping, humanitarian assistance, and peace enforcement within countries, it could be preferable to sta-

---

[2] Figures provided by Headquarters, United States Air Forces in Europe.

tion at least one brigade of U.S. light infantry or air assault forces in Europe at the expense of one of the heavy brigades. A mixed force along these lines would be more deployable than the current structure, capable of a wider range of missions, and would seem to offer more flexibility.[3]

The Army has begun to experiment with the creation of brigade-sized "strike force" units that could be formed in response to an emerging crisis by drawing sub-units from existing forces and coalescing them around a standing headquarters specially configured for this purpose. Perhaps a hybrid type of unit will emerge from these efforts that will be better suited than existing mechanized and armored brigades to the range of contingencies most likely to confront NATO in the coming years. Whatever course this and related efforts take, the watchwords should be rapid deployability and high leverage through the exploitation of information and precision firepower.

Some have called for a review of the U.S. base structure in Europe. After all, if the locus of NATO's most likely and significant contingencies has shifted outward, it seems reasonable that the current base structure should shift accordingly. A closer examination of the basing issue suggests, however, that this question should be approached with care. First, USAF forces based in Europe have already shifted somewhat toward the south and east. Since the end of the Cold War, the U.S. Air Force has stationed two squadrons of multirole F-16 aircraft at Aviano Air Base in Northern Italy. Aviano has been the hub of U.S. and allied air operations over Bosnia. The U.S. Air Force has also been present at Incirlik Air Base in southeastern Turkey since the Gulf War. Forces deployed there today are monitoring and enforcing restrictions on Iraqi military activities in northern Iraq. Given the concurrence of the governments of Italy and Turkey, forces at these bases are capable of conducting operations throughout the central and eastern Mediterranean and parts of the Middle East and North Africa.

---

[3] Proponents of the existing Army force mix in Europe will argue that the mechanized infantry brigades there can "dismount" from their armored vehicles and, thus, replicate the capabilities of a lighter infantry unit. In practice, however, it is rare to find a mechanized infantry unit that is manned to a level that it can put as many soldiers into the field as a comparable light unit. And troops in light units have more time to master small-unit infantry skills and tactics than their brethren in mechanized units.

The utility of bases in the United Kingdom and Germany is less obvious. These bases are far from most potential operating areas, but a host of other factors must be considered as well. One key factor is proximity to training areas. Main operating bases today are best seen not so much as places from which to mount combat operations (their primary role in the Cold War), but rather as places from which to train and to deploy when required. Currently, the best ground force training areas in Europe are in western Germany at Hohenfels and Grafenwoehr. And the United Kingdom offers some of the least-restrictive airspace in Europe for low-level flying training. Units based elsewhere more frequently must "deploy to train." That is, they regularly send personnel and equipment to distant bases for days or weeks at a time to accomplish essential training tasks. This costs valuable time each year that could be spent supporting operations or projecting influence.

Other factors also militate against moving forces to new bases:

- NATO has assured Moscow that it does not intend to station forces from other nations on the territory of its new member states. Doing so would be viewed by many in Russia as a provocation and it is not warranted in the current environment.
- The main bases that U.S. forces in Europe call home are, in general, well-endowed with the facilities needed to support the units stationed there, including their personnel and families. Replicating these facilities elsewhere could cost billions of dollars.
- Less tangible, but also important, is the fact that U.S. forces today are based on the territory of the most important full members of the Alliance—Germany, the United Kingdom, Italy, and Turkey. As such, these forces constitute a unique symbol of our long-standing security ties with these key countries.

In short, the case for keeping U.S. forces (if not the precise units) where they are generally outweighs the arguments for moving them. However, it is critically important that U.S. forces have assured access to quality facilities on the Alliance's eastern and southern fringes. Turkey and Italy emerge as the keys to power projection missions in future zones of instability. In light of this, U.S. planners should be especially attuned to the sensitivities of these host governments regarding the use of their facilities. It is especially important that the United States and Turkey begin now to discuss at senior levels the basis for U.S. Air Force deployments to Turkey following the completion of Operation Northern Watch—the enforcement of restrictions on Iraq north of the 36th parallel. Some diversification of U.S. bases and activities might also be called for. For example, the U.S. and Italian governments might want to begin exploring the desirability of spreading some of the flight activities now conducted at Aviano to facilities located elsewhere in Italy, particularly in the less densely populated south.

Finally, in the wake of NATO's Operation Allied Force, which compelled Serbia's Slobodan Milosevic to withdraw his forces from Kosovo, it is clear that NATO will bear the primary responsibility for enforcing peace in the Balkans for the indefinite future. Without a fundamental change in the nature of the regime in Belgrade, substantial numbers of NATO forces will need to remain in Bosnia-Herzegovina, Kosovo, Macedonia, and perhaps other parts of the Balkans for years to come. The aerial component of this joint and combined force can be based in Italy, Germany, and perhaps other NATO countries, but there will be no substitute for sizable formations of capable ground forces— combat units as well as headquarters, supply, engineering, intelligence, civil affairs, and a host of other critical support functions. This suggests that U.S. and allied planners will need to build permanent facilities in the region for their ground forces, and posture these forces for long-term operations there. Options for the United States include the permanent stationing of a brigade-sized unit in multiple kasserns in Bosnia and Kosovo (along the lines of the Army's deployment in Korea), or rotating units through the region on a regular basis, as is done now in Bosnia.

Chapter 7
The Way Ahead

Many of the trends and challenges addressed in this report involve interests central to the security and well-being of Americans. These include concerns over the future orientation of Russia, energy security, and impeding and coping with the proliferation of WMD. Other interests, such as promoting the spread of democracy and Western values, or preventing interethnic violence, massive abuses of human rights, and the humanitarian crises that accompany such atrocities, although less clear-cut, can nevertheless generate demands for action, including sizable military operations. No nation, including "the world's sole remaining superpower," can secure such interests on its own. By their very nature, the challenges we face demand effective and coordinated action by a large number of states.

The United States, in short, needs partners, not just to share risks and burdens but to provide a basis for effective, multinational responses to events and trends that threaten common interests. In most cases, unilateral American initiatives to address post–Cold War security challenges are not likely to be effective, nor will they often be politically popular at home or welcomed by those abroad at whom they are directed. For their part, our European allies need a secure and stable environment in which their values and way of life may flourish. Achieving this will require active efforts to support reform and modernization throughout Europe and neighboring regions. *In extremis*, military power will be needed to cope with crises and to enforce conformity with minimal norms of state behavior. For the foreseeable future, the United States will remain the only nation capable of conducting large-scale military operations far from its borders. For these reasons, and because the United States and its allies share so many interests and values, an enduring basis exists for a continued transatlantic security partnership.

NATO is the institution best-suited to harmonizing and executing policies on security issues that affect the United States and Europe. Its members share habits of cooperation based on the experience of having worked together for decades to address common security challenges. Moreover, once a consensus is reached at the policy level, NATO's unique system of multinational headquarters, command and control centers, and common doctrine and training allows allied forces to carry out agreed policies in a well-coordinated fashion.

The Alliance is not perfectly adapted to the challenges of this changed and changing world, but it is moving in the right direction. A key factor will be the Europeans' willingness to accept greater responsibility for the defense of common interests outside of the NATO treaty area and, increasingly, beyond Europe itself. A broader conception of NATO's mandate—and the development of the military capabilities needed to support it—will be essential if the members of the Alliance are to succeed for the next 50 years in their stated purposes, including, "to unite their efforts for collective defense and for the preservation of peace and security."

This report is intended to be prescriptive in nature rather than predictive. The assessment of challenges, strategies, and forces offered here points to directions in which the United States and its allies arguably *should* move and is not indicative of where they necessarily *will* go. Each of NATO's member states—including the United States—will need to adopt significant changes in its mind-set and its patterns of defense investment if they are to make this vision of the Alliance as an effective "exporter of stability" a reality. Such changes are never easy, and the difficulties are magnified when no immediate danger looms. Nevertheless, with gradual adaptation and a willingness on both sides of the Atlantic to invest in the military and other capabilities called for to meet emerging challenges, NATO's members stand a good chance of being able to execute their ambitious strategy aimed at bringing peace, stability, and prosperity to a widening circle of nations.